Tiago Goulart
Edelvan S. Claudino
Junio Crevelaro

The future of renewable energies on the European stage

AF294564

Tiago Goulart
Edelvan S. Claudino
Junio Crevelaro

The future of renewable energies on the European stage

Projection for a hypothetical scenario

ScienciaScripts

Imprint
Any brand names and product names mentioned in this book are subject to trademark, brand or patent protection and are trademarks or registered trademarks of their respective holders. The use of brand names, product names, common names, trade names, product descriptions etc. even without a particular marking in this work is in no way to be construed to mean that such names may be regarded as unrestricted in respect of trademark and brand protection legislation and could thus be used by anyone.

Cover image: www.ingimage.com

This book is a translation from the original published under ISBN 978-613-9-60752-5.

Publisher:
Sciencia Scripts
is a trademark of
Dodo Books Indian Ocean Ltd. and OmniScriptum S.R.L publishing group

120 High Road, East Finchley, London, N2 9ED, United Kingdom
Str. Armeneasca 28/1, office 1, Chisinau MD-2012, Republic of Moldova, Europe
Printed at: see last page
ISBN: 978-620-7-30151-5

Index

SUMMARY

Energy developments in Europe from the 1980s to the first decade of the 21st century, looking at the scope of renewable energy generation. Analysing economically, topographically and climatically, in the first contact with the availability of conditions necessary to develop each type of these energies.

The book analyses energy generation data and explains its constant rise due to the advancement of this technology, which is strongly linked to the pillars of sustainability that are also indirectly addressed.

With a holistic approach, it was possible to see how climatic, geological, economic and political constraints impact on sustainable development through renewable energies.

CHAPTER 1

INTRODUCTION

The world has been going through a constant process of warming, which is believed to be caused by emissions of gases called Greenhouse Gases (GHG). With the industrial revolution in the 20th century and the progress in energy generation, which was so necessary for technological development, it was realised that the production and consumption of this energy was, and still is, the main contributing factor to these emissions.

In the midst of this evolutionary scenario, renewable energies have emerged, capable of continuing the world's technological and energy progress and reducing GHG emissions, especially CO_2 (carbon dioxide) and CH4 (methane) from burning fossil fuels (oil).

The largest market for this new pattern of energy production is the developed European market, which has been seeking to exploit its energy potential, especially in the last 20 years, taking into account its geographical characteristics, such as climatic and topographical factors, as well as the technological factors that are more developed on this continent.

Because of Europe's supremacy in this field, the obvious growth of this technology and the environmental importance of the subject, we decided to analyse the development of renewable energies in a select group of 28 European Union countries: Belgium, France, Germany, Italy, Croatia, Luxembourg, the Netherlands, Denmark, Great Britain, Ireland, Portugal, Spain, Greece, Austria, Finland, Sweden, the Czech Republic, Estonia, Hungary, Latvia, Lithuania, Poland, Slovakia, Slovenia, Malta, Cyprus, Bulgaria and Romania.

The aim of analysing the data was to describe the entire process of renewable energy development in Europe, as well as to identify which energy source has been most exploited and is on the rise, thus making it possible to verify the environmental, social and economic nature of these energies, as well as to study a possible investment market and identify trends that could be extended to developing countries such as Brazil, China and India.

CHAPTER 2

THEORISING THE ECONOMIC AND ENVIRONMENTAL ASPECTS

According to the European Parliament, renewable energy sources (wind energy, solar energy, hydroelectric energy, ocean energy, geothermal energy, biomass and biofuels) are alternatives to fossil fuels that help to reduce greenhouse gas emissions, diversify energy supplies and reduce dependence on unviable and volatile fossil fuel markets, particularly those for oil and gas. The EU is a leader in renewable energy technologies. It holds 40 per cent of the world's renewable energy patents and, in 2012, almost half (44 per cent) of global renewable electricity generation capacity (with the exception of hydropower) belonged to the EU. The renewable energy industry in the EU currently employs around 1.2 million people. EU legislation on the promotion of renewable energies has evolved significantly in recent years. The future policy framework for the post-2020 period is being debated.

The UNIDO makes an economic and social assessment of the prospects for each renewable energy analysed, as well as explaining its mitigating nature, as described below:

Small-scale hydroelectric power can be important at a regional level, especially when it is profitable. On the other hand, the construction phase of large hydroelectric plants has social consequences and direct and indirect effects on the environment, such as diverting water, altering hillsides, preparing dams, creating infrastructure for a large labour force, or disturbing aquatic ecosystems, which adversely affect human health. The social consequences include the displacement of people, as well as a boom and bust effect on the national economy. The associated infrastructure stimulates regional development and also represents additional benefits for agriculture such as a water dam.

The technical potential has been estimated at 14,000 TWhe/year, of which 6,000-9,000 TWhe/year can be exploited economically in the long term after taking into account social, environmental, geological and economic factors. The market potential for reducing GHG emissions depends on the fossil fuel replaced by hydroelectric power (IPCC Report, 2007).

Among the possibilities for providing energy is biomass, which includes municipal solid waste, industrial and agricultural waste, existing forests and energy plantations. The yield and costs of biomass energy depend on local conditions, such as the availability of land and biomass waste and production technology.

Normally, the product-input ratio for high quality food crops is reduced when compared to that corresponding to energy crops, which constantly lowers the first ratio by a factor of 10. It is estimated that the cost of biomass production varies greatly. Based on commercial experience in Brazil, it is possible to produce approximately 13 EJ/year of biomass at an average cost of 1.7 $/GJ, in the case of supplying wood waste. Costs are higher in European Union countries. For electricity generation in these countries, it is believed that biomass inputs in the future will cost around 2 $/GJ.

The variation in mitigation costs for forms of energy derived from biomass, such as electricity, heat, biogas or transport fuels, not only depends on the cost of biomass production, but also on the economic aspects of certain fuel conversion technologies.

Considering biomass costs of 2 $/GJ and small-scale production, it would be possible to generate electricity at 10-15 0/kWhe. With a lower biomass cost (0.85 $/GJ), it would be possible to generate electricity at less than 10 0/kWhe. By replacing coal with biomass, mitigation costs would range from 200 to 400 $/t C avoided. In a future integrated biomass gasifier/gas turbine cycle with an expected efficiency of 40-45% and biomass costs of 2 $/GJ, it would be possible to produce electricity at a cost comparable to coal and/or coal prices in the range of 1.4-1.7 $/GJ. In this case, the specific mitigation costs could be negligible.

Modern biofuels derived from wood raw materials offer the possibility of producing more energy, at lower cost, with lower impacts on the environment than those produced by most traditional biofuels. In addition to ethanol, methanol and hydrogen are promising fuels.

Currently, modern biomass conversion technologies and biomass plantations are in their infancy and require more research to reach technical maturity and be economically viable. With concerns about future food supplies, the issue has arisen that in African and other underdeveloped countries, there will be no land for biomass production for energy purposes. Possible competition for land use will depend on the degree to which agriculture can be modernised in these countries to achieve yields equivalent to those obtained by European Union countries, and on whether agricultural production is intensified in an ecologically and economically acceptable way.

Wind energy is another great renewable energy possibility and in a large network it can contribute approximately 15-20% of annual electricity production, without special provisions for storage, reserve or load management. In a public system with a predominance of fossil fuels, the mitigation effect of wind technologies corresponds to a reduction in the use of fossil fuels. Wind potential in 2020 is expected to be in the 700-1000 TWhe range; if it were used to replace fossil fuels, without taking costs into account, this would translate into a reduction in CO2 emissions of 0.1-0.2 Gt C/year. The average cost of current stocks of wind energy is around \$10/kWh, although the range is wide. Costs could be lower for large wind farms.

In countries with a large number of wind turbines in operation, there is sometimes opposition from the population due to factors such as turbine noise, the visual effects on the landscape and the disturbance of wildlife.

Direct conversion of sunlight into electricity and heat can be achieved through photovoltaic (PV) technology and solar thermal energy. PV energy is already competitive as an independent energy source away from electric utility grids. However, it is not competitive in most grid connection applications. Although modular capital costs have fallen a lot in recent years, system capital costs are 7,000-10,000 \$/kW; (2,400 kWh/m^2/year). However, the costs of PV systems are expected to improve considerably through research and economies of scale. Because of its modularity, PV technology can reduce costs through experimentation and technological innovation. Although PV devices do not contaminate in normal operation, in some systems toxic materials have to be used, so there can be risks in the manufacturing, use and disposal phases.

Between 2020 and 2025, the annual economic potential of solar energy in small, well-defined markets has been estimated at 16-22 EJ. Realising this potential will depend on improvements in the cost and performance of solar thermal energy technologies.

If this potential is fully realised, regardless of costs, the CO_2 reduction could be 0.3-0.4 GtC per year. The cost of mitigation in relation to a coal-based electricity generation of 5 0/kWh would be in the range of 130170 \$/t C avoided; compared to gas-based electricity with similar costs, this range would be 270-350 \$/t C avoided. These costs take into account aspects of the energy system, such as storage needs or the advantages of substituting more expensive electricity during periods of high load, when PV production is closely related to maximum electricity demand.

Solar thermal electrical systems can meet a considerable part of the world's electricity and energy needs in the long term. With this technology, heat is generated at high temperatures, so that a conversion efficiency of around 30 per cent can be achieved.

There are various emissions associated with geothermal energy, including CO_2, hydrogen sulphide and mercury. Advanced technologies are closed-circuit and their emissions are very low. It is estimated that from 2020 to 2025 the potential for geothermal energy will be 4 EJ. Deep hot dry rock and other non-hydrothermal reserves offer new supply resources. Despite their importance to the global economy, the possibilities for carbon reduction are scarce.

Although the total energy of tidal flows, waves, thermal gradients and salinity in the world's oceans is large, it is likely that over the next 100 years only a small part of it will be exploited, and technological change can be stimulated because each of them offers a continuous incentive for research and development of emission reduction technologies in order to avoid the tax. On the other hand, the purchase of emission reduction quotas is uncertain. This situation is reversed in the case of emission quotas.

CHAPTER 3

HOW AND WHY STUDY THIS SCENARIO?

Looking at the world panorama of the evolution of renewable energies, we can easily see that the European continent stands out in the production of this type of alternative energy generation, thus being chosen as a more effective and diversified object of study for analysing the data.

Due to the difficulty of collecting data directly, which would have had to include travelling to the European region, data was collected from the European Energy Commission, as well as other websites that helped with the theoretical and even practical background, which made this book possible.

The analysis itself was based on a natural tendency, where it is observed that an evolution of the energy sector is extremely necessary to meet technological and population growth, factors that potentially increase the demand for energy, especially in countries with a high level of development.

CHAPTER 4

A MATHEMATICAL VIEW OF RENEWABLE ENERGIES

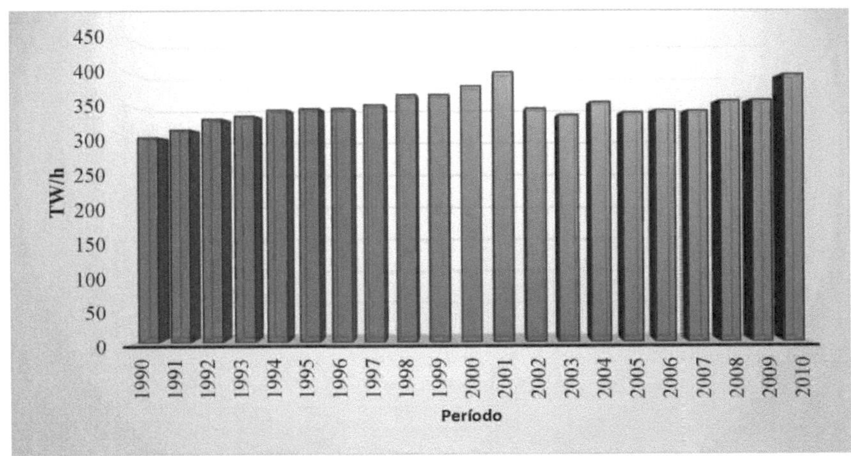

Figure 1 - hydroelectric power production graph

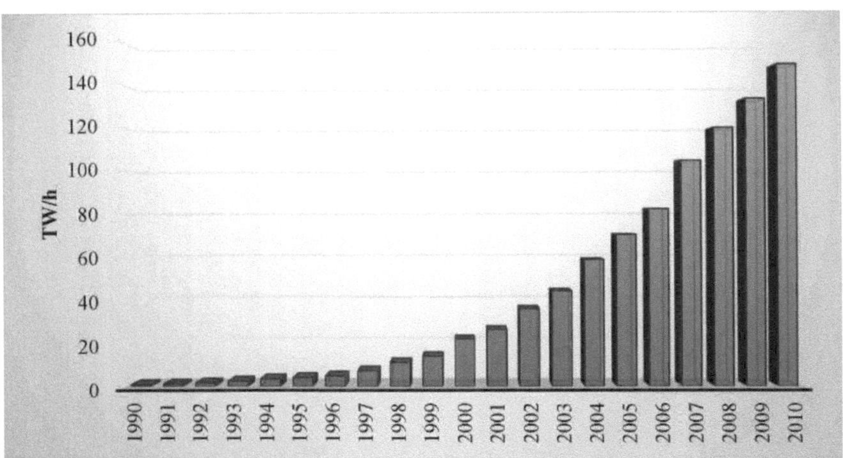

Figure 2 - wind energy production graph

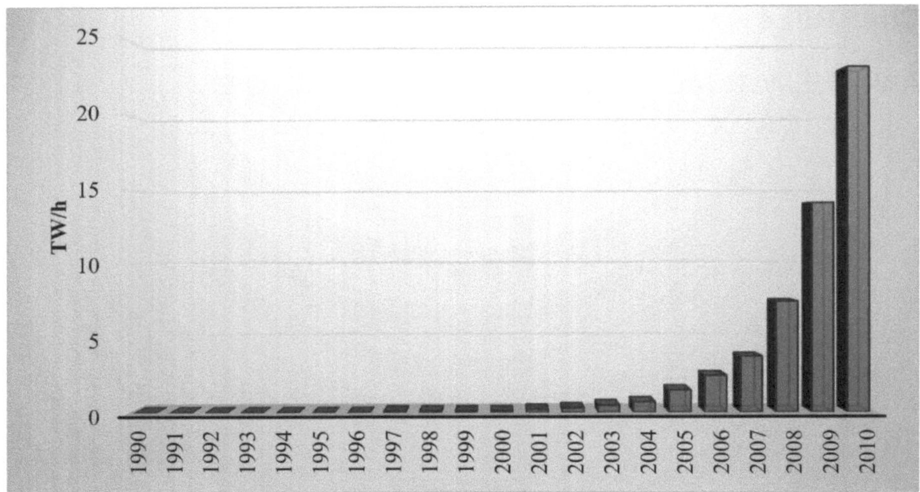

Figure 3 - solar energy production graph

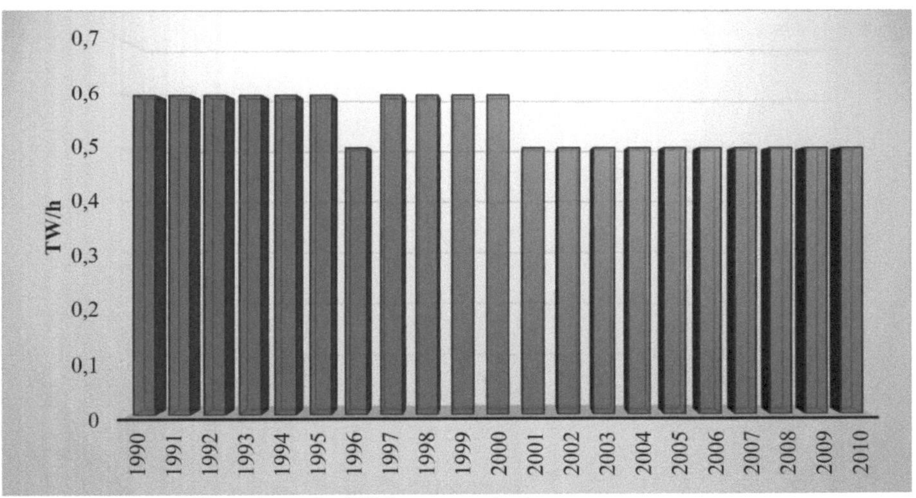

Figure 4 - graph of tidal energy production

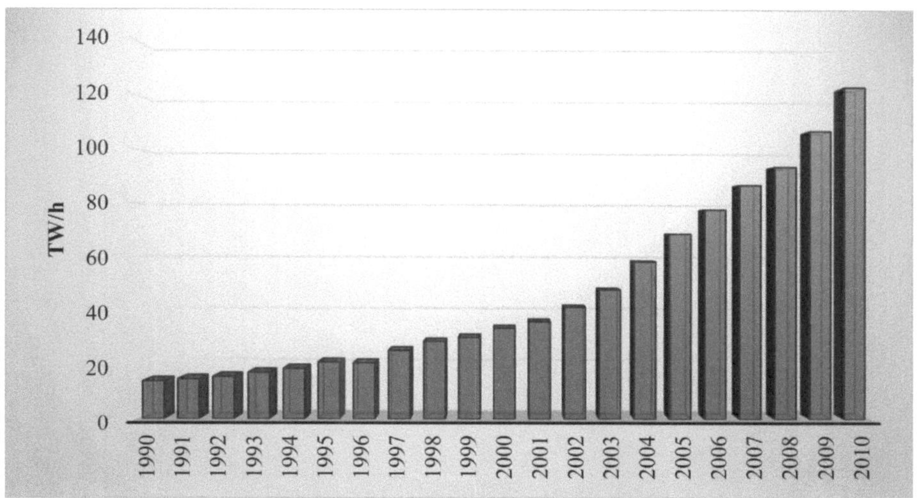

Figure 5 - biomass energy production graph

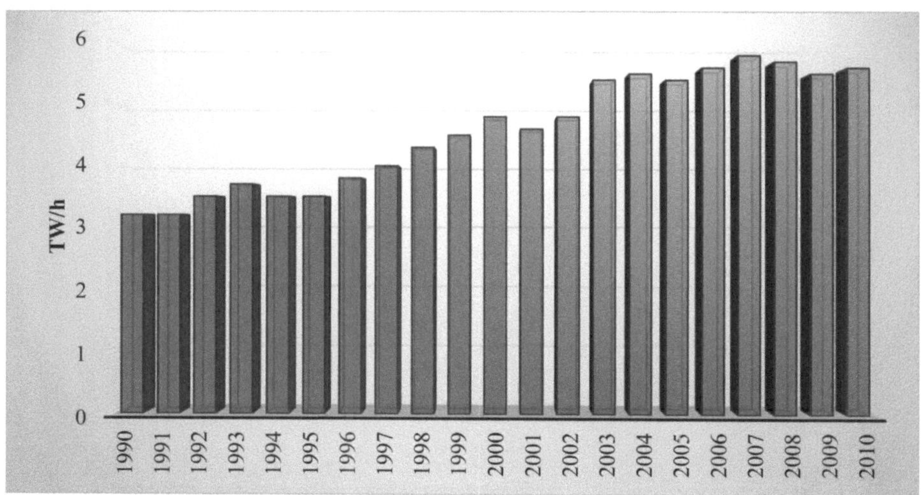

Figure 6 - geothermal energy production graph

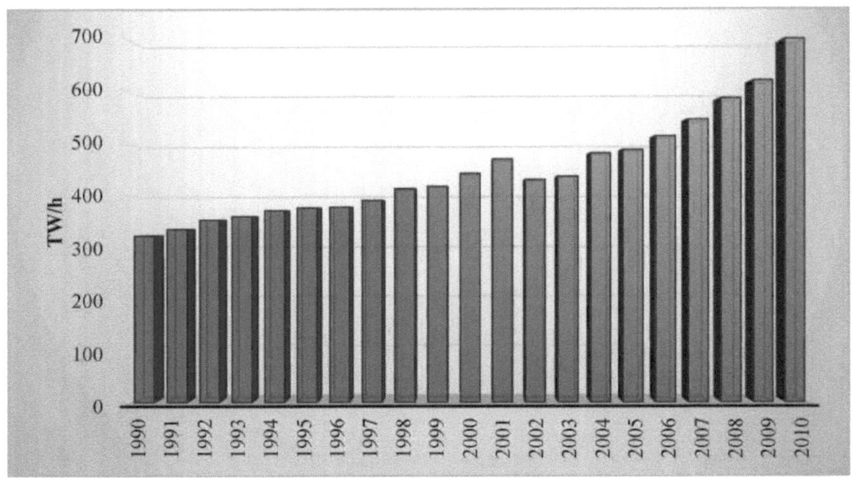

Figure 7 - graph of total renewable energy production

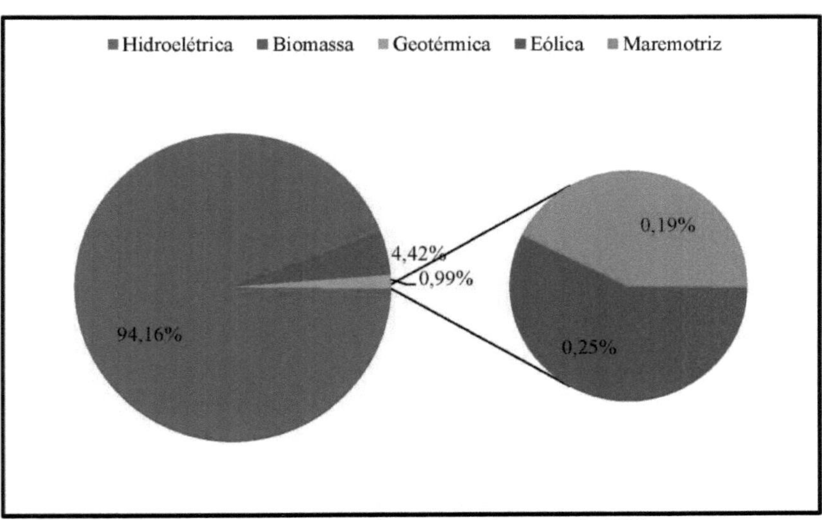

Figure 8 - Graph of electricity production from renewable sources in 1990.

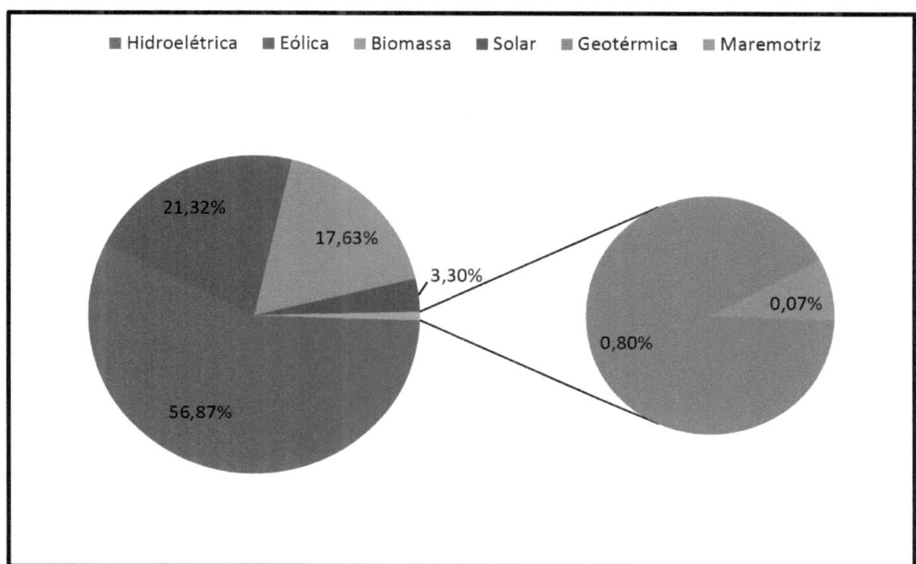

Figure 9 - Graph of electricity production from renewable sources in 2010.

CHAPTER 5

ANALYSING REGRESSION

In order to describe energy production in Europe over time using a regression model, we have:

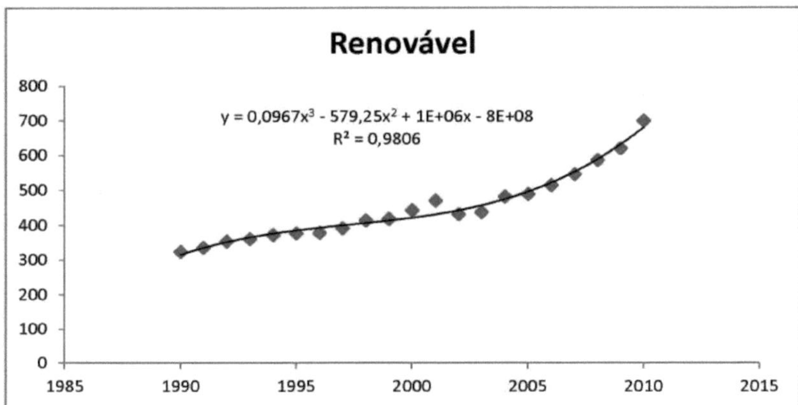

Figure 10 - regression of renewable energy as a whole

Analysing the trend of the graph, we see a cubic trend that shows the possibility of growth in the near future, and the graph has an excellent R^2 , equal to 98%.

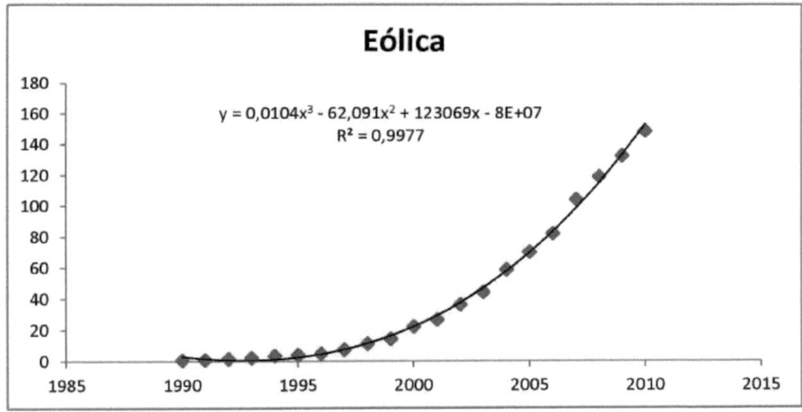

Figure 11 - wind energy regression

Analysing the trend of the graph, we see a cubic trend, and the graph has an excellent R^2 , equal to 99 %.

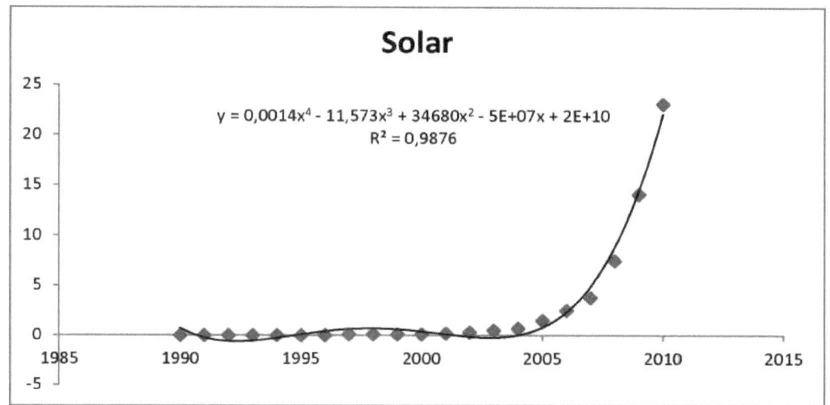

Figure 12 - wind energy regression

Analysing the trend of the graph, we see a fourth-degree trend, and the graph has an excellent R2 of 98 %.

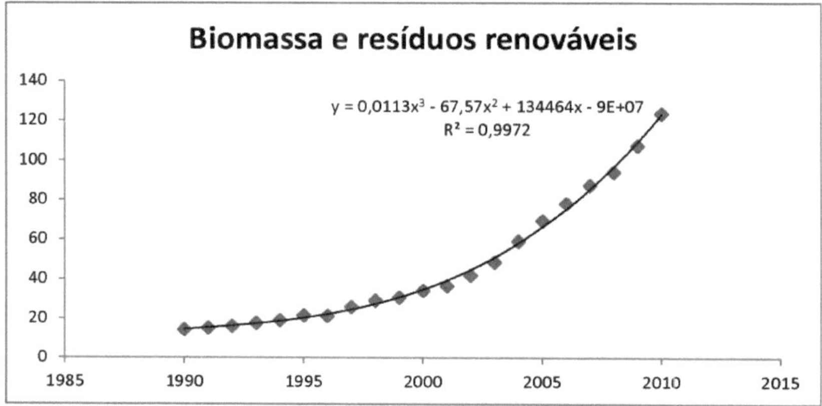

Figure 13 - wind energy regression

Analysing the trend of the graph, we see a cubic trend, and the graph has an excellent R^2 , equal to 99 %.

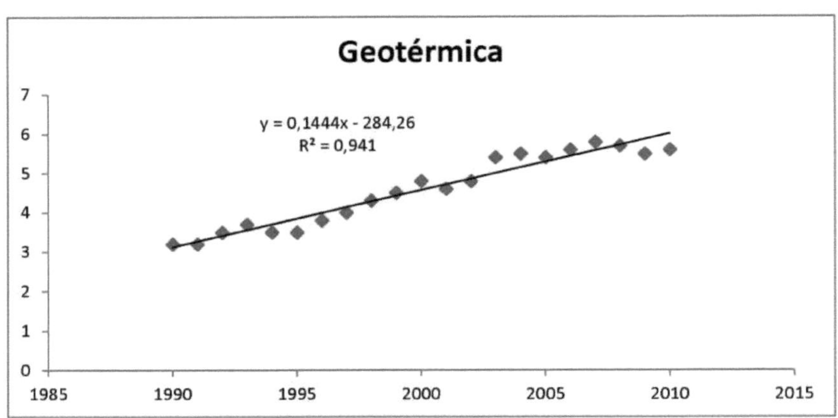

Figure 14 - wind energy regression

Analysing the trend of the graph, we see a linear trend, and the graph also has an excellent R2, equal to 94 %.

Based on a direct analysis of the graphs presented and the information added up about the European continent and its economic development in recent years, we can safely observe that the trend in the development of renewable energies is one of gradual and constant growth, meeting the continent's energy demand, which will follow this same evolutionary trend, but with some reservations regarding the individuality of each energy presented:

Hydroelectric power: This is directly responsible for the variation in the total energy graph. It represents great energy potential as well as considerable and decisive economic potential. In the last year studied, it made up more than 56 per cent of Europe's total renewable energy production, and peaked in 2001 when it accounted for an impressive 86 per cent of clean energy produced and 401.5 TWh generated. As we have seen, after the year of its peak in energy production, hydroelectric power showed a decline that can be economically and socially explained by the attack on the twin towers on 11 September 2001 in the USA.

The terrorist attack that cost the Americans a fortune, around 3.3 trillion dollars, according to an analysis by the national newspaper The New York Times, triggered a global economic crisis that had a direct impact on the European continent. Not only did the country with the world's largest economy have to stop investing in certain energies that were being developed on the old continent, but it also had to take out loans with Europe, which at the end of the same year would face further economic problems with the implementation of the euro as a unified currency and the so-called response to the "War on Terror", which directed a significant part of the countries' economies towards protection against further terrorist attacks. This period of financial recession was reflected in a drop in total and hydroelectric power generation, especially in 2002, after which the graph rose again to keep up with the economic and energy demand needed for European development.

Wind, solar and biomass energy: After the signing of the Kyoto Protocol and especially ECO 92, the world began to worry more effectively about environmental issues and as a consequence began to invest more in alternative and clean energy sources. This led to a greater development of these types of energy during the first decade of the 21st century, which followed a natural trend of gradual and constant growth in their productive potential.

Tidal and geothermal energies: These two energies are the ones that show the greatest fluctuation in their production levels, both of which, in addition tc also suffering from the effects of economic contractions, are relatively new and little explored compared to the others mentioned above.

Tidal energy depends a lot on climatic factors that allow it to be better captured, and may not be directly related in the short term to the investments made in it. On the other hand, geothermal energy, which is presented as the best natural way to generate energy, has not yet been fully "accepted", mainly because of the noise it generates, with some countries assuming a certain hegemony in the generation of this type of energy, which happens, for example, in Iceland in the case of Europeans.

An analysis of the graphs in Figures 8 and 9 shows the natural tendency for the renewable energy matrix to be more heterogeneous, where even with the constant growth of hydroelectric power it has gone from around 94% to around 56.87%, which translates into faster growth for most of the other clean technologies. It was also possible to analyse that wind energy is the fastest-growing, with electricity generated by wind energy increasing by 186 times over the course of 20 years, an occurrence that can be explained in part by the fact that wind energy can produce 12% of the world's energy demand and avoid the emission of 10 billion tonnes of CO2 in 12 years, according to the Global Wind Energy Outlook 2008 report, drawn up in partnership by the Global Wind Energy Council (GWEC) and Greenpeace. With wind energy's great electrical potential raising its economic importance and its ability to reduce emissions being so compelling, any country that has the necessary climatic conditions should invest in this type of energy, which is considered infinite.

With this in mind, developed and emerging countries have begun to develop this technology. Among the top 10 countries in wind energy in 2009, according to a study by Allianz, are India and China representing the emerging countries, the USA representing the American continent and the other 8 belonging to the group of 27 European countries chosen as the focus of study for this article (Germany, Italy, Denmark, France, the United Kingdom, Spain and Portugal). Brazil, which only had its first aero-generator installed in 1992 in Fernando de Noronha, already occupies 21st place in the same ranking where China and the USA lead. The country has received incentives from the BNDES as well as foreign investment in this sector, which promises to put Brazil on the list of the world's top 10 wind energy producers in just two years, according to the Energy Research Company (EPE).

CHAPTER 6

COMPARATIVE PROJECTION FOR THE YEAR 2040

In order to make more distant projections in the time frame, and from that point on to develop economic theories that more closely match reality, it is necessary to have coherent factors related to purely mathematical issues, such as analysis of climatological and geological factors, population behaviour, timing and regional economic outlook, among others. For the purposes of this study, we have assessed some of these factors in more depth and others have been taken into account in a less comprehensive way, but they have not been left unanalysed. In the projection that follows, some energies have been favoured for analysis, once again taking into account points of greater importance for a broader and more significant evaluation, such as: greater advances in productivity and expansion, easier dissemination throughout the continent, better utility as a comparison between energies and the change in the pattern of their use, among other points studied for the analysis. Energy from the following sources was studied with greater emphasis: Biomass and waste renewables, Geothermal, Solar, Wind and Petrolium products.

The table below shows the formulae that have made it possible to find mathematical values for the development of energy generation through the sources researched, using the factors already discussed and which will be detailed in greater detail later. These formulae may differ from those found earlier in this same study, since the formulas shown above, as well as making a shorter-term projection, are geared towards a purely mathematical perspective where all that really mattered was how good the coefficient of determination (R^2) calculated along with the formula was, since it represents the measure of fit for the model found.

Projection formulae used		
Geothermal	$f(x) = 0.1444x - 284.26$	R2 = 0,941
Solar	$f(x) = 12954*\ln(x) - 98507$	R2 = 0,967
Oil derivatives	$f(x) = -7.5842x + 15345$	R2 = 0,913
Biomass	$f(x) = 10046*\ln(x) - 76309$	R2 =

		0,876
	f(x) = 14330*ln(x) -	R2 =
Wind power	108881	0,845

Table 1 - mathematical projection formulas

By analysing the energies separately, and taking into account the diversity of topics to be considered when choosing the best mathematical projection models for the chosen year of 2040, we can see, at first glance, that some models are mathematically adjusted satisfactorily while others don't follow the approximations very well, resulting in the coefficient of determination being far from its perfect measure, which would be $R^2 = 1$. However, as we have already seen, we are not only interested in the linear efficiency of the projections, which makes us turn our attention to a wider range of mathematical influencers.

The energy derived from the geothermal source has a linear factor that did not change abruptly in the analysis up to 2040, which allowed us to use the same function to find its respective electricity generation potential for this period. Among the renewable energies analysed at this stage of the study, geothermal energy has a relatively low figure, both in terms of current and future generation, even though it has managed to develop considerably in terms of percentage, as we will see later, this is due to two main reasons, firstly because of its location in Europe, this continent is not the most favourable for the evolution of this type of energy compared to others, even if we currently have some projects under development with this source in particular, it will still be a little used source, and the second reason is linked to the non-exploration and development of technologies aimed at this area in a totalitarian way, some countries like Switzerland and Germany can make Europe look like a leading continent in this type of energy generation, but it is a punctual generation that in order to understand the European scenario in its entirety, does not add much.

Another energy that stands out for its coefficient of determination is solar energy. Previously we had seen that the function of this energy behaved like a fourth degree equation, but considering an analysis for the year 2040 we realised that this behaviour of the curve would be far removed from reality, since as much as energy is capable of evolving and increasing the capacity of the energy matrix, it will still behave in a logarithmic way when factors that limit its growth are taken into account, such as periods of financial crisis. The renewable energy sector, despite proving to be an economical and more sustainable solution, still suffers from a process of evolutionary braking every time countries have to go through times of recession, and solar energy is no exception.

Of particular note in our study, due to the environmental gains this brings, is the linear decline of oil-based products. After dominating the energy scene for a long time, there is a natural tendency for this energy source to be gradually replaced, until its contribution to the energy matrix is zero. It is believed that the automotive sector will be the last to abandon the use of oil as its main energy generator, but Europe is still ahead of other continents, firstly because it has only Russia as an oil producer in the list of the world's top ten, and secondly because it is a pioneering continent in the use of hybrid forms of energy generation for transport.

It is intuitive that energy demand tends to grow year on year. Taking this into account, once we have stipulated the trends in the occupation of these renewable energies in the energy matrix, we can make a numerical analysis stipulating how much each energy will produce, in TWh (Terawatt-hours), for the year 2040, thus facilitating a correlative analysis of investment possibilities in these areas, as well as population growth and technological advances, factors that up until now have been related to the growth in global energy demand.

Growth forecast for 2040		
	2015	2040
Geothermal	6,7	10,32
Solar	51,88	211,61
Oil derivatives	62,8	0
Biomass	129,7	267,7
Wind power	147	323,7

Table 2 - Growth projection

Looking at table 2, one can conclude that there has been a massive advance and occupation of Europe's energy matrix by sources considered clean, which according to the forecast, would lead to a gradual reduction in the space filled by oil sources, making derivatives of these sources almost unusable for energy generation.

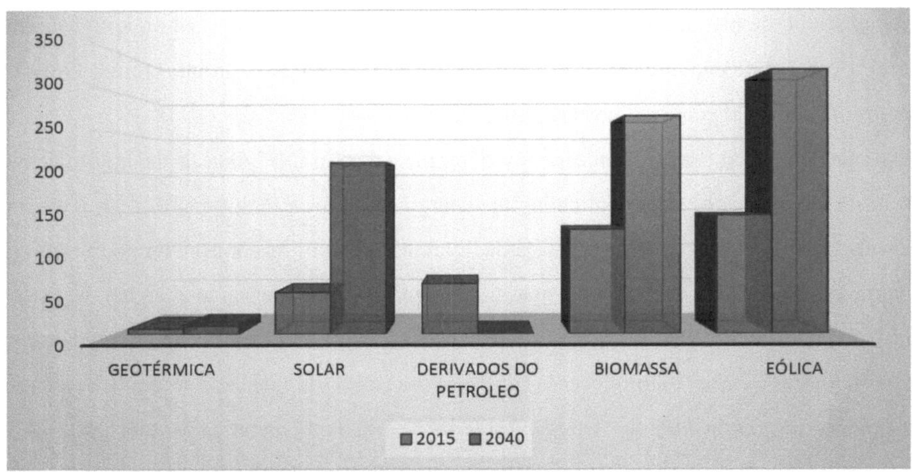

Figure 15 - numerical comparison chart between 2015 and 2040

Energy sources from biomass and wind (wind) play an important role in changing the energy landscape. They already represent good productivity and the future scenario shows even greater growth in these energies. Like all the others, in addition to having impacts such as the production of certain gases (Biomass) and noise and visual pollution (Wind), and some other factors such as the need for technological improvement and the construction and occupation of large spaces (wind farms), they reveal a favourable relationship when analysing the cost-benefit factor.

Socially, we can think of factors that would lead some countries to be more reluctant to replace oil-based energy sources with renewable sources. One of the most discussed factors in this regard is undoubtedly the workers who work in, for example, coal mining industries, and who would lose their jobs if this source were to be deactivated as a primary energy source. In addition, switching from one energy development pattern to another is costly and burdensome in many ways. Starting from these negative points, we can open up a broader discussion.

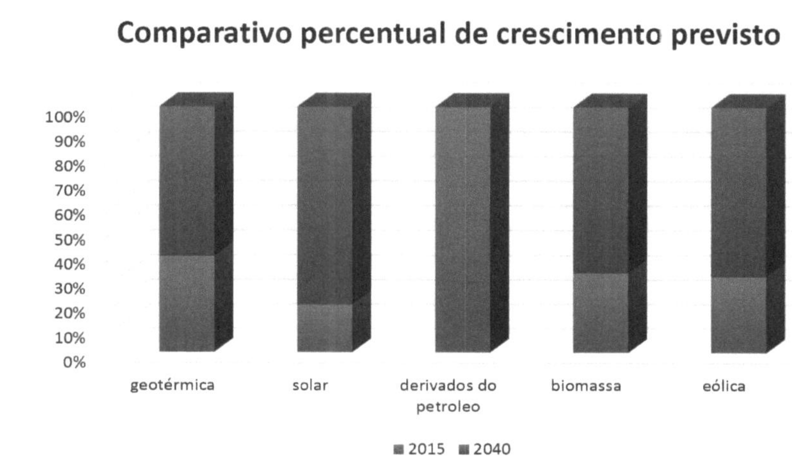

Figure 16 - percentage comparison chart between 2015 and 2040

With regard to the increase in the unemployment rate, you have to think on a macro scale, where the introduction of new technologies is also an introduction of new labour markets, so the jobs that are, in theory, lost with the replacement of fossil fuels will also be replaced by jobs generated through renewable energies. From an economic point of view,

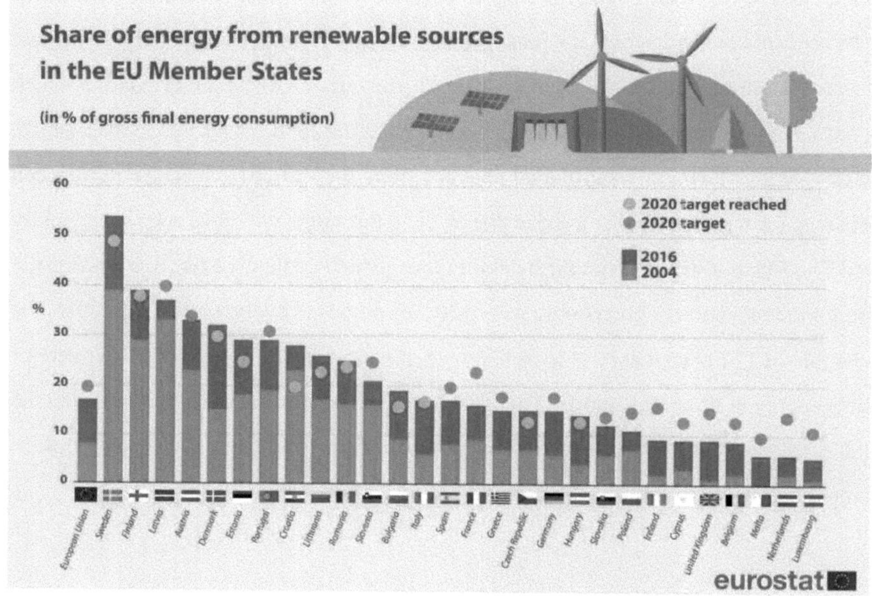

Figure 17 - graph of the percentage of renewable energy used in the matrix

you have to think about how much your country can still develop with backward technologies that, due to global trends, will become more expensive and worthless within a few years. Renewable energies show progress in the three pillars that represent sustainability. And the importance given by the main EU countries to sustainable growth can be seen in the objectives that have been set, both for reducing emissions, which is the consequence, and for implementing new technologies that are considered clean in the energy generation process, which is the cause.

Bearing in mind that consumption patterns needed to change, the European Union participated in the two phases of the Kyoto Protocol, before and after it was ratified: the first (essentially for industrialised countries), which provides for a 5% reduction in emission levels by 2012, with 1990 as the reference year, and the second, which provides for an 18% reduction in these levels by 2020. As we can see from the graph in figure 17, which shows the gross percentage of renewable energy in the energy matrix, it is clear that some countries have already managed to achieve their 2020 targets in 2016, while other countries are still a fair way off achieving this. France is a negative highlight, needing to reach the target of 23 per cent, in 2016 it was still at 16 per cent, and with a higher annual growth rate of just 0.9 per cent, which at this rate would make it impossible for the country to reach the target. On the other hand, Sweden stands out positively, with a target of 49% set for 2020. In 2016 it reached 53.8%, which, despite not having shown any substantial changes from 2015 to 2016, has already reached and surpassed the target set, so the trend is that by 2020 this target will have been successfully achieved.The EU as a whole is not far behind France, which could give rise to a pessimistic scenario for the reduction of greenhouse gas emissions proposed by the protocol. In 2016, the EU had a rate of 17 per cent and, as is well known, it has set itself the goal of 20 per cent of renewable energy in its energy matrix. This goal is fundamental to the objective set out in the protocol, since the change in the source of energy used is precisely to produce energy that, in addition to other benefits, reduces GHG emissions.

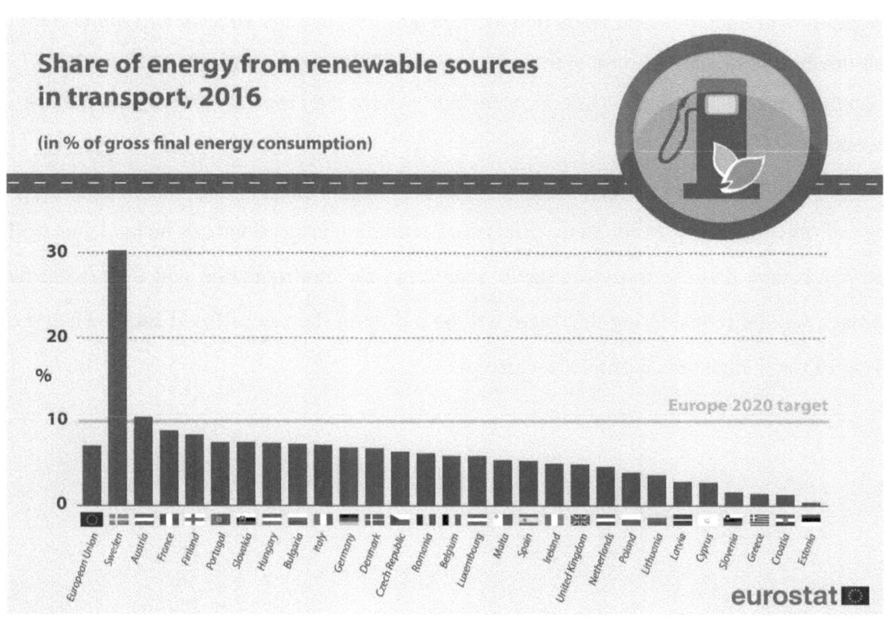

Figure 17 - graph of the percentage of renewable energy used in the matrix
Source: Eurostat 2018

you have to think about how much your country can still develop with backward

technologies that, due to global trends, will become more expensive and worthless within a

few years.

Renewable energies show progress in the three pillars that represent sustainability. And the

importance given by the main EU countries to sustainable growth can be seen in the

objectives that have been set, both for reducing emissions, which is the consequence, and

for implementing new technologies that are considered clean in the energy generation

process, which is the cause.

Bearing in mind that the pattern of consumption needed to change, the European Union

participated in both phases of the Kyoto Protocol, before and after it was ratified: the first

(essentially for industrialised countries), which provides for a 5% reduction in emission

levels by 2012, taking 1990 as the reference year, and the second, which provides for an

18% reduction in these levels by 2020. As we can see from the graph in figure 17, which

shows the gross percentage in a broad and holistic way, we must include energy developments in each of these fields and increasingly improve the sectors that are ahead on the development scale. Some of the countries highlighted can offer a kind of model, both on the positive side and on the negative side, where they represent models not to be followed.

Cross-referencing the data collected by the study here with the projections for 2040 opens up a range of other possibilities for study. Emissions reduction projections can be made, as well as the percentage that the transport sector represents for this reduction and even what its percentage will be (considering that there will be a drop in the use of fossil fuels) when the last year of the mathematical projection arrives.

CHAPTER 7

ANALYSING CLIMATIC FACTORS

When it comes to renewable energy, there are several analytical factors that contribute to its mass production, one of which is the climatological study of an area or region. With this in mind, Europe signed a treaty in 2009 called the *Climate and Energy Package,* one of the objectives of which is to combat climate change by limiting the rise in the average global temperature to 2 °C, based on the principle of reducing greenhouse gas emissions by 20%. This commitment was reinforced in 2015 by the Paris agreement, which limited the temperature rise to 1.5 °C, so that the climate issue would not jeopardise the production of renewable energy, since various energy sources depend on this factor, including, for example, solar energy, wind energy and biomass.

For a climatological analysis of renewable energies, it is essential to understand the minimum requirements for the production of each one, so it is important to obtain the necessary data and interpret it in the best way.

IRRADIATION AND SOLAR ENERGY

In the case of solar energy, the climatological part is based on the principle of solar irradiation on an inclined plane, since the inclination is equal to the local latitude, so that the maximum amount of solar energy is captured without depending on the topography of the area (studied and explained elsewhere in the book). Figure 19 shows the average annual solar irradiation map of the European continent.

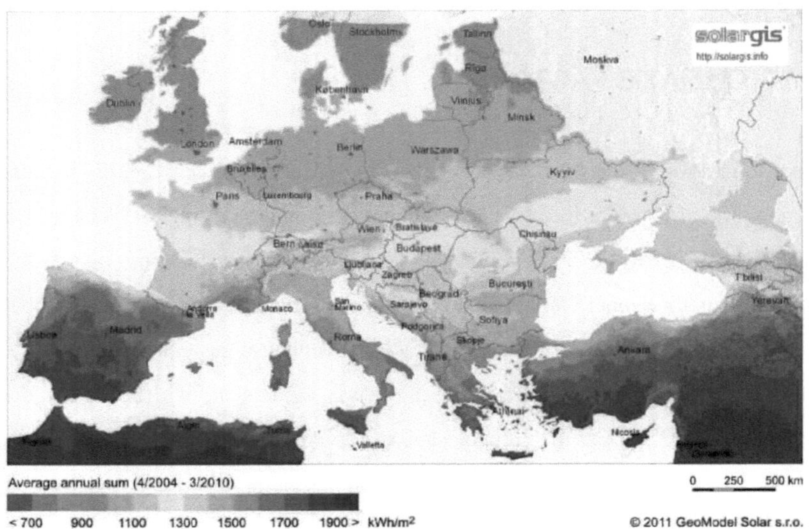

Figure 19 - Map of average annual solar irradiation on the European continent, in kWh/m²

Based on the analysis of figure 19, a photovoltaic map drawn up to show the total annual solar irradiation on the European continent, the colours represented by the legend show solar irradiation levels in [kWh/m2], where the values vary between 700 and 1900 kWh/m2. Considering the irradiation conditions shown on the map, it can be seen that the countries in the redder colours would have greater potential for solar energy production, since these countries would have irradiation levels varying between 1300 and 1900 kWh/m2. However, the reality is different, since according to data from the European Parliament, the country with the highest solar energy production on the European continent is Germany, which has lower irradiation levels than other countries such as Italy and Spain, for example. Figure 20 below compares solar energy production levels in [GW] in 2015.

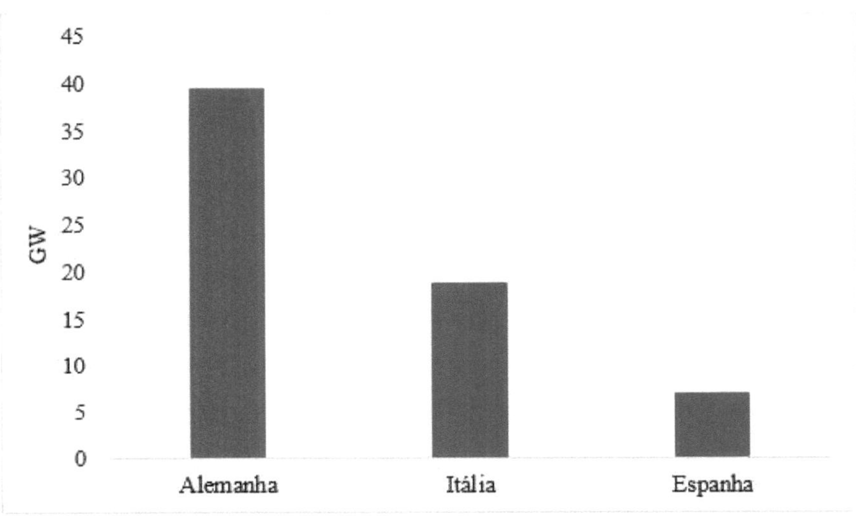

Figure 20 - Top solar energy producers in 2015

According to the data presented in figure 20, it shows that even though Germany does not have high levels of irradiation, it is the leading producer of solar energy on the European continent. This is due to the country's political investments, since the German government uses the strategy of competition between renewable energies and established technologies. These political strategies have made Germany the country that generates the most solar energy *per capita* and the second largest producer of solar energy, behind China. This political plan becomes more evident when analysing solar energy per working day, which meets a third of the country's demand.

CHAPTER 8

WIND ENERGY AND THE INCIDENCE OF WIND

Climatological studies are also necessary to understand the production of wind energy and its characteristics, knowing that winds originate due to the different heating of the earth's surface, i.e. the unequal incidence of the sun. In order to use it, it is necessary to convert the kinetic energy of translation into the kinetic energy of rotation of the aero-generators to generate electricity (as mentioned in the chapter *analysing regression*). Wind energy is measured and quantified by means of speed sensors by institutions around the world, and wind speed is generally measured in [m/s]. Figure 21 below shows data from the wind map of the European continent, a study carried out by the *Risoe National Laboratory*.

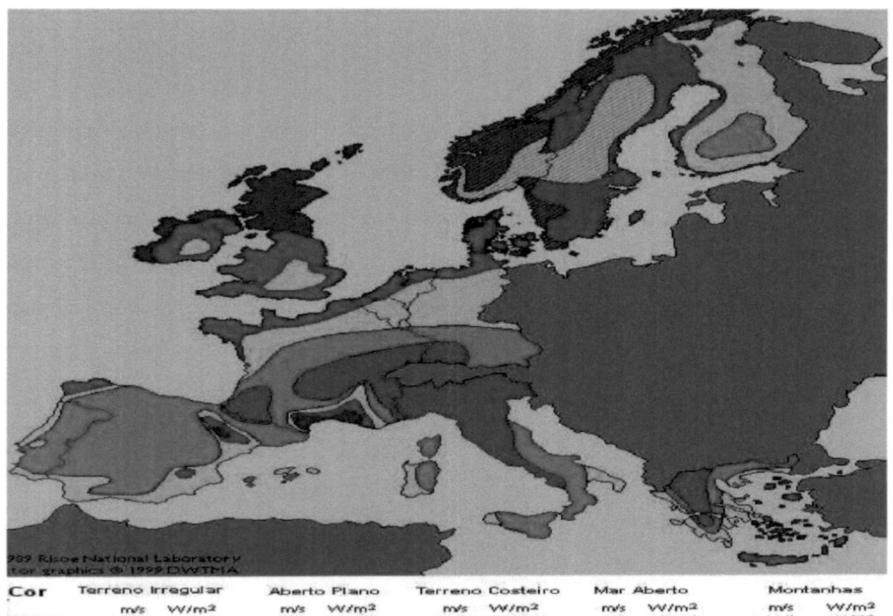

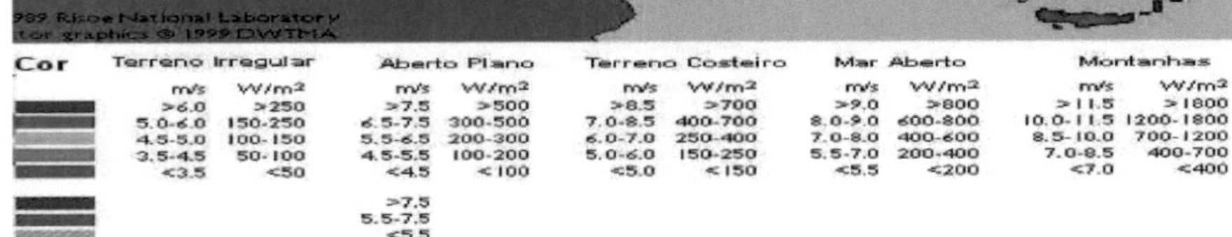

Figure 21 - Wind map of the European continent

Evaluating the data proposed in figure 21, and through knowledge of the legend, shows that the countries to the south of the continent have a greater tendency to produce wind energy, since countries like Norway, England and Sweden have wind speeds of between 5 and 10 [m/s], with variations over the seasons. However, due to political factors and investments in the domestic market, Germany has once again taken first place as the country with the highest wind energy production in Europe, even though it has lower wind speeds of around 3.5 to 5 [m/s]. Figure 22 below compares the wind power potential already produced in 2015 between the countries with the highest wind power trends and Germany. Although there are no climatological requirements that contribute to the production of wind power in its favour, the German government, through a policy called *Energiewende,* intends to make a transition in energy sources, leaving aside fossil fuels and promoting renewable energy sources. All of this has helped Germany stand out in terms of renewable energy sources, as the incentives provided by the national government have made it easier and cheaper to produce energy.

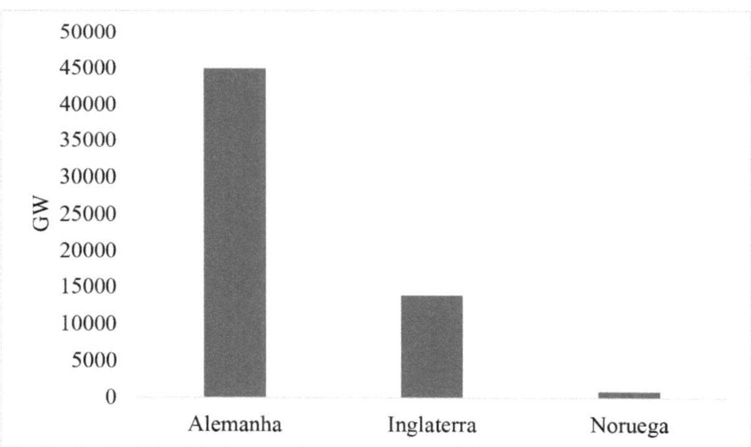

Figure 22 - Countries with the highest wind trends x Germany

CHAPTER 9

AVAILABILITY X BIOMASS ENERGY PRODUCTION

We can also relate climatological factors to biomass production by considering the total amount of energy generated from this source (solid, liquid and gaseous biomass) and relating the generation capacity to agricultural and forestry biomass, the more traditional means, which comes from agricultural waste (and livestock through biogas) and wood subgroups. Climate has a direct influence on vegetation growth conditions and agricultural potential, indirectly altering the availability of biomass for energy production. Therefore, by observing the European biomes and their plant and mainly tree composition, as well as the agricultural activities already mentioned in the discussion on land use and capacity, we can conduct a broad assessment of the local conditions for developing energy through this source.

Figure 23 - the planet's biomes
Source: Biology - César and Sezar

The map in figure 23 shows the world's main biomes, among which we can see the predominance of temperate rainforests, chaparral and taiga in Europe. Of these, temperate rainforests have denser, more arboreal vegetation and are therefore considered a better source of resources for biomass energy production than Taigas and Chaparral (which have vegetation that is arboreal but more spaced out and shrubby, respectively). In addition, temperate forests have a greater biodiversity of plant species, which favours mixed uses in search of greater calorific potential.

Another environment that offers biomass energy a high potential for development is agriculture and the production of energy from its waste.

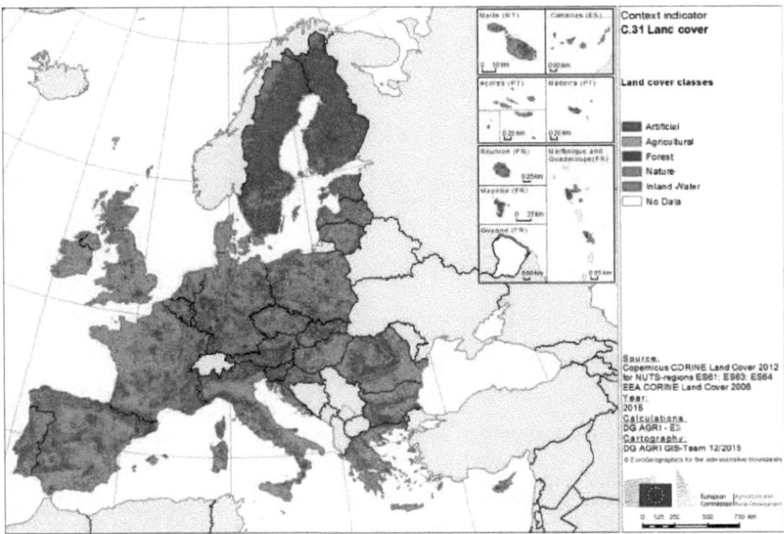

Figure 24 - European land cover
Source: European Commission for Agriculture and Rural Development

Analysing figure 24, it is possible to see a predominance of agricultural land, which, when combined with forest and pasture land, occupies around 85% of the European Union's territory. This factor highlights the potential that this region has for producing energy through biomass, since pasture areas make biogas generation from animal and forest sources favourable, while agricultural waste generates raw material for the production of biofuels and thermal energy from burning solid plant biomass.

Countries such as Germany, France, Italy, Spain, the Netherlands and Belgium are prominent on the European and world stage in the production and export of agricultural products, reaching a percentage of approximately 30 per cent of all world agricultural exports in 2002. But as with other energies, we can highlight which of these countries are actually taking advantage of these favourable conditions to generate energy from the biomass in their territories.

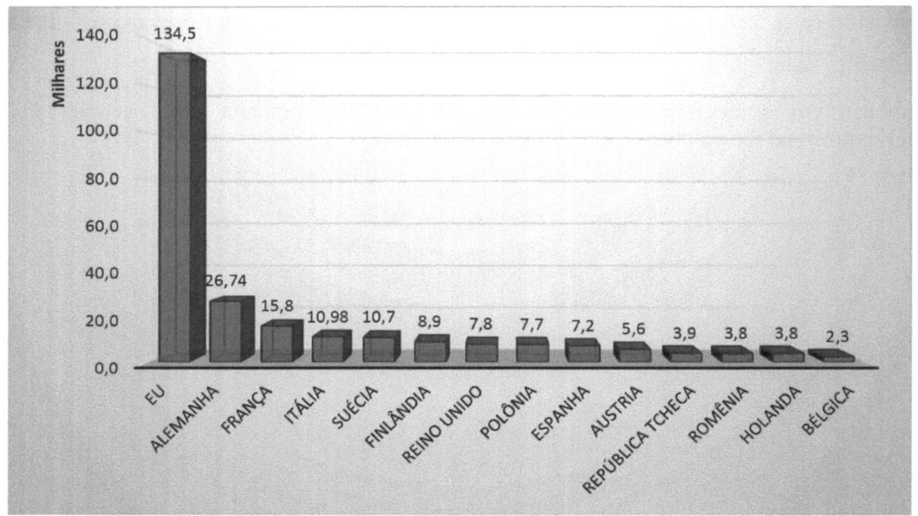

Figure 25: Graph of biomass production in millions of TOEs in 2016
Source: adapted from Eurostat 2018

By following the performance of the countries mentioned above, we can see that Germany, France and Italy stand out in this sector, taking advantage of the climatic and geographical conditions that are favourable to them. Germany alone accounts for 19.88% of the entire European Union's production, followed by France (11.75%) and Italy (8.16%). We can consider that due to a lack of investment and technology, countries such as Spain, Holland and Belgium (which together account for 9.81%), despite having the raw materials needed to develop this type of energy, do not exploit it to its full potential, maintaining a low level of production of this type of energy. This factor makes it clear that our entire discussion is pertinent when we consider environmental factors in energy production, but it is worth emphasising that political, social and economic facts must always be taken into account, as was done in our projection.

CHAPTER 10

GEOLOGICAL FACTORS: RENEWABLE ENERGY AND ITS USE
SOIL

European soil is the basis for 90 % of all food, feed, fibre, fuel and energy production, providing raw materials for activities ranging from horticulture to the construction sector. Soil is also essential for the health of the ecosystem: it purifies and regulates water, maintains nutrient cycles and supports biodiversity, thus playing an important role in the possible mitigation of climate change and its impacts.

However, even though European society is aware of this, it makes excessive demands on the soil by supplying it with large continuous loads.

 As a result, the soil's capacity to provide ecosystem services in terms of food production, biodiversity reserves and the soil's function as a regulator of gases, water and nutrients is under pressure.

The observed rates of sealing, erosion, decline in organic matter and contamination in the soil reduce its resistance and its capacity to absorb the changes to which it is subjected, currently generally being used as illustrated in Figure 26.

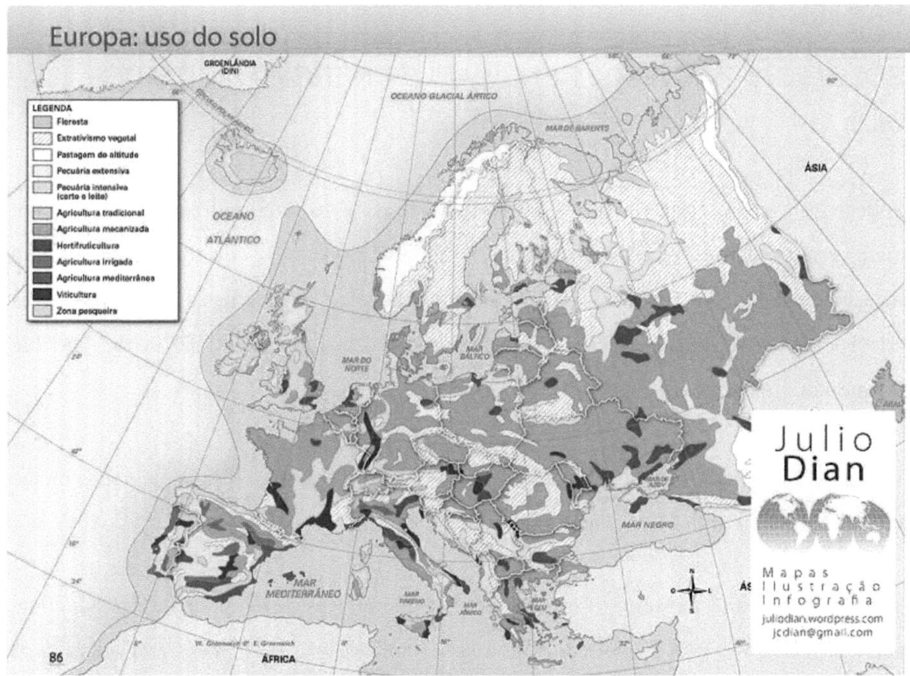

Figure 26 - European land use map.
Source: Current Geography, 2018.

We can analyse that a large part of land use on the European continent is due to the use and cultivation of agriculture. We know that with the exacerbated growth of the population and consequently of the consumable goods provided by the use of the soil, it becomes something with a shorter useful life and has medium-term future possibilities of ceasing to be a renewable source, which it would be if it were used correctly, even in an ecosystem with exhaustible resources.

There are ways of effectively interrelating soil and its use with renewable energies, analysing its geomorphological and pedological conditions for the installation and generation of these energies. It is of great value and knowledge to know and explain these energies, such as geothermal, wind, biomass, photovoltaic and oil-derived energies, which we know from mathematical estimates can reach close to zero due to their non-renewable generation sources.

The presence of geothermal heating in Europe has already been perfected in some regions that have shown and are showing interest in investing in this energy, since, as already mentioned, it comes from regions rich in volcanic soils and also from the extraction of the heat provided by deep rocks (5000 m) that consist of water reserves heated to 200°C. One of the first European countries to invest in this method due to the factors that contribute to its generation was Italy, specifically in the region of Tuscany, where in mid-1908 volcanic soils were discovered that heated the underground water, resulting in the first geothermal power plant built in the world, supplying the villages of the city of Larderello in its entirety, and today there are 35 geothermal power plants in the region.

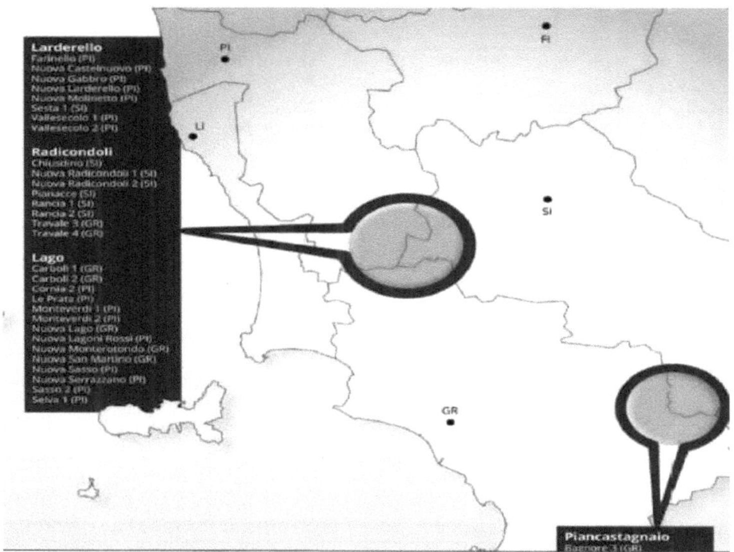

Figure 27 - Geothermal power plants in Tuscany.
Source: ARPAT - Tuscany Regional Agency for Environmental Protection.

Switzerland has a recent history of investing in areas with great geothermal potential, but its initial project was unsuccessful, specifically in 2007 in Basel when attempts to drill without a primary study and local reconnaissance caused seismic tremors in the region, causing the project to be cancelled momentarily. However, around 50 geothermal installations have sprung up in Switzerland (Figure 28), supplying homes, industries and hotels, making it the country with the highest density of this type of project in the world, but without any geothermal power plants.

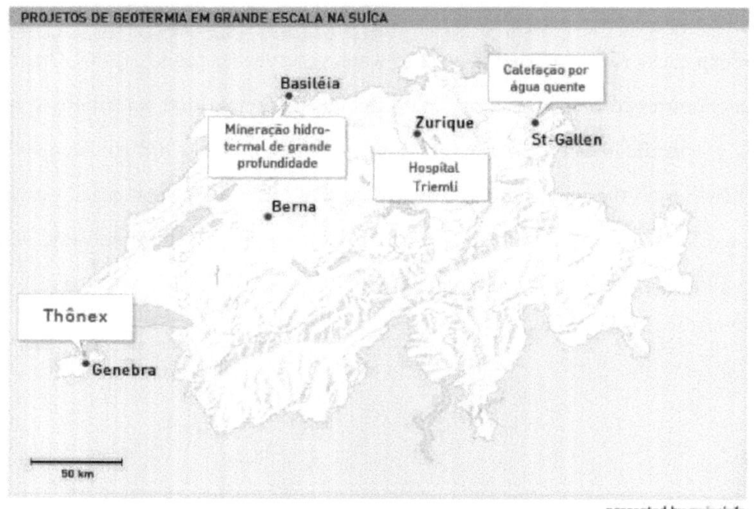

Figure 28 - Geothermal installations in Switzerland.
Source: Swissinfo, 2018.

Given that there are no geothermal power plants in the country, the Geneva project was approved. By 2020, more than 10,000 homes in Thônex will be receiving natural heat and electricity from deep underground, with excavations of 5,000 metres in rocky soils in search of geothermal sources.

As it is a region of diverse soils with the existence of Podzol, soils rich in iron and lime, which is the reason for its name (*pod* = under/under, *zola* = grey/ash), it becomes a typical soil for the development of the *Taiga and* the *Boreal Forest*, enabling Biomass to be generated and used as a source of energy for the population, as this soil is fertile and widely used for planting trees that provide wood and also pasture.

Not only has this region contributed to the production of biomass, but Europe as a whole has great potential in its different forms, from sticks to pellets to wood sawdust (or, to use the fashionable name, biomass), which accounts for around half of Europe's renewable energy consumption. With wood, Poland and Finland account for more than 80 per cent of the demand for renewable energy. Even though Germany is a major investor in wind and sunlight as an energy generator, with large subsidies directed towards them, 38 per cent of non-fossil fuel consumption comes from wood, making it directly and indirectly beneficial for the regional, national and international population and economy, with both gains and losses (Figure 29).

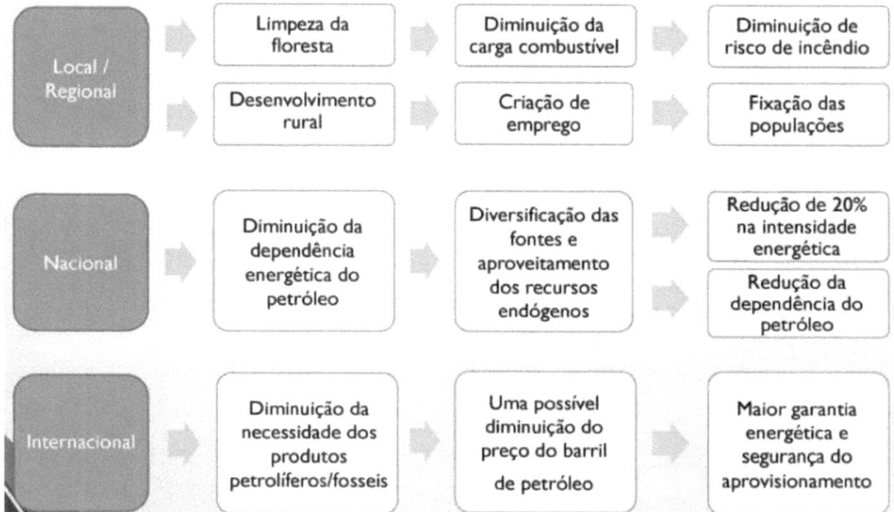

Figure 29 - Advantages and implications of using biomass for energy production.
Source: International Biomass Congress, 2017.

With limited sources of raw materials, the UK, Belgium and the Netherlands have been relying on imports of industrial wood pallets to reduce the use of coal in some of their energy utilities. Although investment in biomass is increasing, the greater the subsidies, the greater the production and emission of CO_2 (Figure 30). Once planted, it is necessary to produce the pellets, which in their process generate large quantities of carbon dioxide compared to what will be produced with the product generated and realising that the greater the production, the greater the deforestation, this soil can become infertile after a certain period of time.

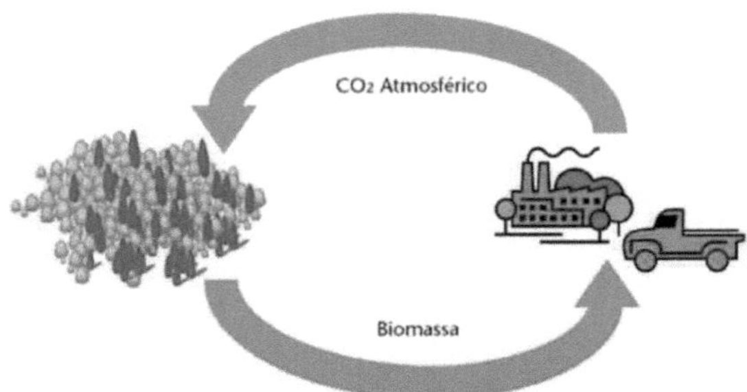

Figure 30 - co2 generation in biomass production.
Source: Renewable Energy Portal, 2018.

A European survey indicated that due to the high consumption, countries on other continents are already being studied that have the potential to supply biomass, a fact that further highlights the exacerbated use of European soil. Australia, South Africa and Brazil stand out among these alternative countries.

Germany is working effectively and in the opposite direction to the major European powers when it comes to renewable energies, as the country is investing heavily, and not just today, in wind energy. Places with significant elevations and even plains have been favourable locations for implementing this energy technique.

Not only Germany has collaborated with land use, but also France, the Netherlands and the Netherlands (Belgium) have invested in this field due to their geographical location (Figure 31), as they are also located on a characteristic soil classified as *Loess* (Figure 32), a very fertile soil with a yellow colour, It's a very fertile yellow-coloured soil, which, in addition to carrying nutrients and seeds, is capable of moving smaller grains of sandy soil, which encourages agricultural cultivation and the installation of large wind turbines. These regions have a good slope, which helps with these types of projects.

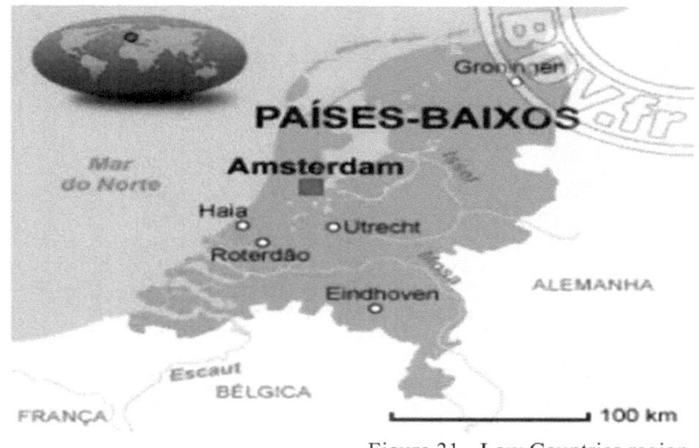

Figure 31 - Low Countries region.
Source: Google, 2018.

Figure 32 - *Loess* soil.
Source: Geography of Europe, 2013.

Solar energy also benefits from regions that have extensive areas with flat characteristics, regions that are also heavily used in Germany to exploit solar energy. Germany is the first in Europe in terms of generation, along with Italy, which uses Cambissolo (Figure 33) (strongly, to imperfectly drained, shallow to deep, bruno or bruno-yellowish in colour, and with high to low base saturation and chemical activity of the colloidal fraction), due to its geological and pedological characteristics, and because of this the region contributes to high investment in this energy.

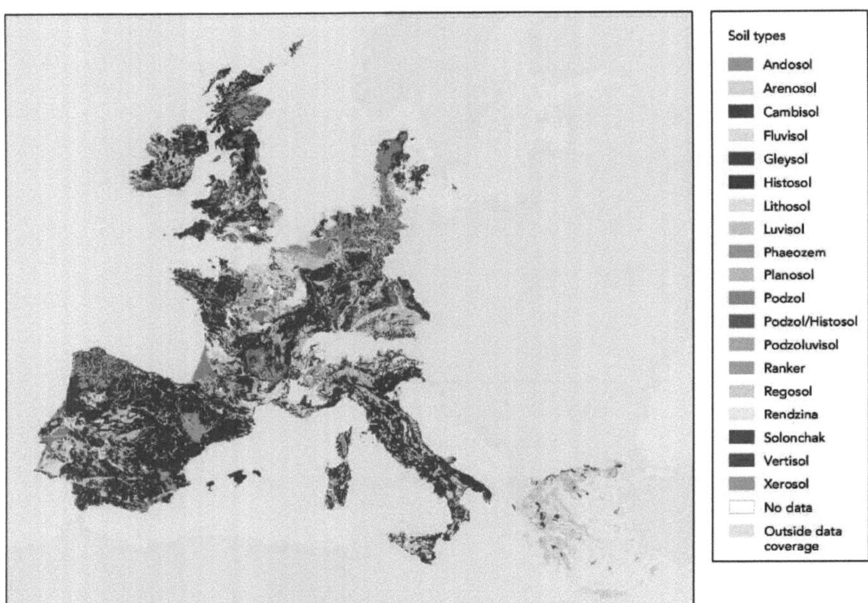

Figure 33 - Soil map of Europe.
Source: Europa Environment Agency, 2003.

Throughout Europe there are different types of soils and characteristics. Developing studies that focus on making better use of these soils can contribute more effectively to the continent's renewable energy potential, enabling these environments to be favourable for the production of raw materials (in the case of biomass) and the installation of solar, wind and geothermal structures.

CHAPTER 11

Final considerations

As a result, it is clear that the European and global energy matrix will continue to receive percentage increases in electricity generation from the huge wind turbines that are yet to be installed, a factor that contributes not only to the economic issue, but why not, above all, to the environmental issue. It's also clear that wind energy has already become a solid and growing market that can generate future gains not only for countries that are already developing this energy efficiently, but also for countries that are still looking to develop it more effectively, such as Brazil. Other energies can also complement this increasingly common scenario around the world, especially in Europe.

Solar energy is still a pillar to be explored, where the availability of solar irradiation is often not converted into energy generation. Countries such as Germany, Italy and Spain (respectively) stand out in the production of such energy in Europe, but Germany, which is first in generation and distribution, is not even close to being the country that receives the most irradiation from the sun, due to other factors already mentioned.

In addition to these important and well-established energies, it is worth highlighting the current growth and projected future growth of biomass. This is a considerably new technology that still has many untapped resources available. Once again, the highlight in this sector goes to Germany, which in terms of renewable energy generation has contributed greatly to the European Union's figures and its objectives.

The one point that stands out from this scenario, but which is fundamental to closing this equation, is the decline in the production of oil derivatives, which after dominating the energy scene for many years, are now losing ground to clean energy sources. The forecast is that by 2040 their consumption in the European Union will fall to a percentage rate very close to zero, which undoubtedly represents a huge environmental gain and, with the emergence of a new market, also an economic gain.

BIBLIOGRAPHY

EUROPEAN COMMISSION. - Country Factsheets, 2012. 169 p.

http://ec.europa.eu/energy/observatory/countries/countries_en.html

https://ec.europa.eu/clima/policies/international/negotiations_en

https://europa.eu/european-union/topics/transport

http://pt.euronews.com/2018/02/06/rumo-a-transicao-energetica-os-casos-sueco-e- Polish

http://www.energyplan.eu/smartenergyeurope/

UNIDO - Renewable Energy Observatory for Latin America and the Caribbean - training programme in renewable energy.

IPCC - Intergovernmental Panel on Climate Change

Minitab 16 and Action - Student version

EUROPEAN COMMISSION.

https://europa.eu/european-union/topics/energy_pt

Accessed on: March 2018

EUROSTAT

http://ec.europa.eu/eurostat/statistics-explained/index.php/Energy_statistics_introduced#Main_issues

Accessed on: April 2018

METEOBLUE

https://www.meteoblue.com/pt/tempo/mapa/wind950hpa/europa

Accessed on: April 2018

ÉPOCA MAGAZINE https://epoca.globo.com/ciencia-e-meio-ambiente/blog-do-planeta/noticia/2017/06/aposta-da-alemanha-em-energia-solar.html
Accessed on: April 2018

TIEPOLO, Gerson Máximo et al. **COMPARISON BETWEEN PHOTOVOLTAIC GENERATION POTENTIAL IN THE STATE OF PARANÁ WITH GERMANY,**

ITALY AND SPAIN. 1. ed. [S.l.: s.n.], 2014. 9 p. Available at:
<https://www.researchgate.net/profile/Gerson_Tiepolo/publication/275828922_COM
PARAGRAPH_between_the_potential_for_photovoltaic_generation_in_the_state_of_paran
a_and_almania_italy_and_spain/links/5547f2
930cf2e2031b384c36/COMPARAGRAPH_between_the_potential_for_photovoltaic_gener
ation_in_the_state_of_parana_and_almania_italy_and_spain
FOTOVOLTAICA-NO-ESTADO-DO-PARANA-COM-ALEMANHA-ITALIA-E-
ESPANHA.pdf>. Accessed on: 02 Apr. 2018.

European Environment Agency.
https://www.eea.europa.eu/pt/themes/landuse/intro, 13 April 2018

Regional Agency for Environmental Protection in Tuscany,
http://www.arpat.toscana.it/temi-ambientali/sistemi-produttivi/impianti-di- produçãoione-
di-energia/geotermia/energia-geotermica , 13 April 18

International Biomass Congress.
http://slideplayer.com.br/slide/12139229/br/slide/12139229/ , 13 April 18

Renewable energy portal.
http://energiasrenovaveis.com/DetalheConceitos.asp?ID content=1&ID area=2&I D sub
area=2,13 April 18

Regione Toscana, geothermal energy, http://www.regione.toscana.it/-/geotermia , 13
April 18

Geography in Europe, https://pt.slideshare.net/carlosribeiromedeiros/geografia-da- europe-
geography-physics-pedology , 13 April 18.

Madeira Magazine, Issue 136.
http: //www.remade .com.br/br/revistadamadeira materia.php?num=1684&subj ect=Bi
omassa&title=Madeira%20%E9%20%20a%20major%20fonte%20de%20energia%2
0renov%E 1vel%20da%20Europa , 13 April 18.

Interest in Geothermal Energy is growing, https://www.swissinfo.ch/por/cresce-o-
interest-in-geothermal-energy/7114776,13 April 18.

Energy Portal, https://www.portal-energia.com/o-mito-da-energia-eolica-na- europa/ , 13

April 18.

Embrapa Technological Information Agency,

http: //www.agencia. cnptia.embrapa.br/gestor/bioma caatinga/arvore/CONT000g798r

t3o02wx5ok0wtedt3n5ubswf.html , 13 April 18.

Euronews, Which country leads in wind energy in the European Union?

http://pt.euronews.com/2017/02/16/qual-pais-lidera-a-energia-eolica-na-uniao-

Europe, 13 April 18.

I want morebooks!

Buy your books fast and straightforward online - at one of world's fastest growing online book stores! Environmentally sound due to Print-on-Demand technologies.

Buy your books online at
www.morebooks.shop

Kaufen Sie Ihre Bücher schnell und unkompliziert online – auf einer der am schnellsten wachsenden Buchhandelsplattformen weltweit! Dank Print-On-Demand umwelt- und ressourcenschonend produziert.

Bücher schneller online kaufen
www.morebooks.shop

Printed by Books on Demand GmbH, Norderstedt / Germany